Ben Stacy Jerrik (Ed.)

Selma, Indiana

Ben Stacy Jerrik (Ed.)

Selma, Indiana

Liberty Township, Delaware County

Part Press

Imprint

Permission is granted to copy, distribute and/or modify this document under the terms of the GNU Free Documentation License, Version 1.2 or any later version published by the Free Software Foundation; with no Invariant Sections, with the Front-Cover Texts, and with the Back- Cover Texts. A copy of the license is included in the section entitled "GNU Free Documentation License".

All parts of this book are extracted from Wikipedia, the free encyclopedia (www.wikipedia.org).

You can get detailed informations about the authors of this collection of articles at the end of this book. The editors (Ed.) of this book are no authors. They have not modified or extended the original texts.

Pictures published in this book can be under different licences than the GNU Free Documentation License. You can get detailed informations about the authors and licences of pictures at the end of this book.

The content of this book was generated collaboratively by volunteers. Please be advised that nothing found here has necessarily been reviewed by people with the expertise required to provide you with complete, accurate or reliable information. Some information in this book maybe misleading or wrong. The Publisher does not guarantee the validity of the information found here. If you need specific advice (f.e. in fields of medical, legal, financial, or risk management questions) please contact a professional who is licensed or knowledgeable in that area.

Any brand names and product names mentioned in this book are subject to trademark, brand or patent protection and are trademarks or registered trademarks of their respective holders. The use of brand names, product names, common names, trade names, product descriptions etc. even without a particular marking in this works is in no way to be construed to mean that such names may be regarded as unrestricted in respect of trademark and brand protection legislation and could thus be used by anyone.

Cover image: www.ingimage.com
Concerning the licence of the cover image please contact ingimage.

Publisher:
Part Press is a trademark of
International Book Market Service Ltd., 17 Rue Meldrum, Beau Bassin, 1713-01 Mauritius
Email: info@bookmarketservice.com
Website: www.bookmarketservice.com

Published in 2011

Printed in: U.S.A., U.K., Germany. This book was not produced in Mauritius.

ISBN: 978-613-8-64691-4

Contents

Selma, Indiana

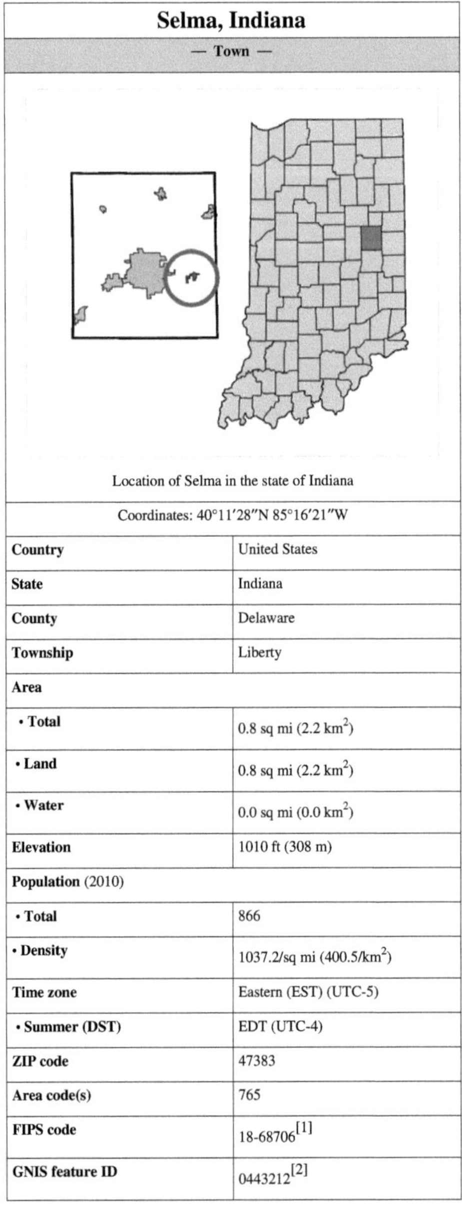

Selma, Indiana

— Town —

Location of Selma in the state of Indiana

Coordinates: 40°11′28″N 85°16′21″W

Country	United States
State	Indiana
County	Delaware
Township	Liberty
Area	
• **Total**	0.8 sq mi (2.2 km^2)
• **Land**	0.8 sq mi (2.2 km^2)
• **Water**	0.0 sq mi (0.0 km^2)
Elevation	1010 ft (308 m)
Population (2010)	
• **Total**	866
• **Density**	1037.2/sq mi (400.5/km^2)
Time zone	Eastern (EST) (UTC-5)
• **Summer (DST)**	EDT (UTC-4)
ZIP code	47383
Area code(s)	765
FIPS code	18-68706[1]
GNIS feature ID	0443212[2]

Selma is a town located in Liberty Township, Delaware County, Indiana. As of the 2010 census, the town had a total population of 866. It is part of the Muncie, IN Metropolitan Statistical Area.

Geography

Selma is located at 40°11'28" North, 85°16'21" West (40.191147, -85.272440)[3] . Indiana State Road 32 runs along the northern edge of the town.

According to the United States Census Bureau, the town has a total area of 0.9 square miles (2.3 km^2), of which, 0.9 square miles (2.3 km^2) of it is land and none of it is covered by water.

Demographics

As of the census[1] of 2000, there were 880 people, 336 households, and 250 families residing in the town. The population density was 1,037.2 people per square mile (399.7/km²). There were 349 housing units at an average density of 411.4 per square mile (158.5/km²). The racial makeup of the town was 97.84% White, 0.57% African American, 0.34% Native American, 0.00% Asian, 0.00% Pacific Islander, 0.11% from other races, and 1.14% from two or more races. 0.34% of the population were Hispanic or Latino of any race.

There were 336 households out of which 40.2% had children under the age of 18 living with them, 59.2% were married couples living together, 13.1% had a female householder with no husband present, and 25.3% were non-families. 22.0% of all households were made up of individuals and 10.7% had someone living alone who was 65 years of age or older. The average household size was 2.59 and the average family size was 3.00.

In the town the population was spread out with 28.4% under the age of 18, 7.8% from 18 to 24, 31.7% from 25 to 44, 19.3% from 45 to 64, and 12.7% who were 65 years of age or older. The median age was 34 years. For every 100 females there were 88.8 males. For every 100 females age 18 and over, there were 86.9 males.

The median income for a household in the town was $44,423, and the median income for a family was $50,357. Males had a median income of $35,333 versus $23,625 for females. The per capita income for the town was $18,361. 4.7% of the population and 4.5% of families were below the poverty line. Out of the total population, 7.1% of those under the age of 18 and 3.8% of those 65 and older were living below the poverty line.

References

[1] "American FactFinder" (http://factfinder.census.gov). United States Census Bureau. . Retrieved 2008-01-31.

[2] "US Board on Geographic Names" (http://geonames.usgs.gov). United States Geological Survey. 2007-10-25. . Retrieved 2008-01-31.

[3] "US Gazetteer files: 2010, 2000, and 1990" (http://www.census.gov/geo/www/gazetteer/gazette.html). United States Census Bureau. 2011-02-12. . Retrieved 2011-04-23.

Liberty Township, Delaware County, Indiana

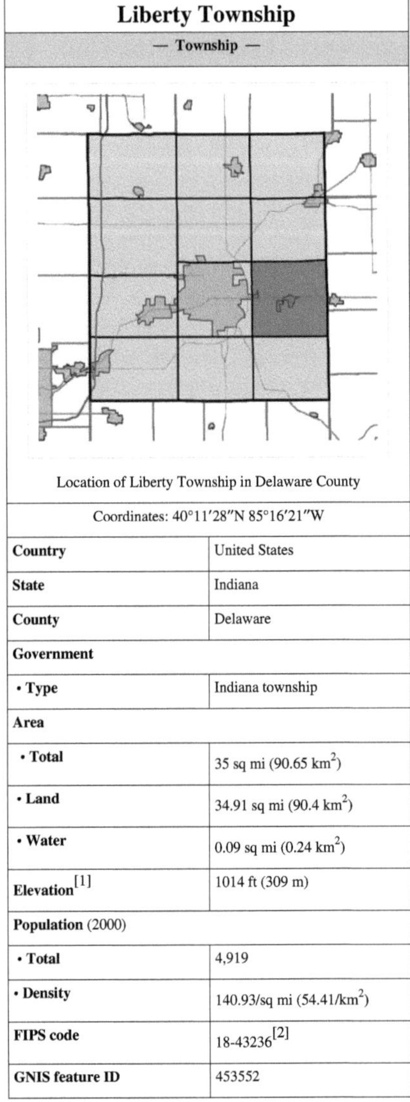

<table>
<tr><td colspan="2" align="center">Liberty Township
— Township —</td></tr>
<tr><td colspan="2" align="center">Location of Liberty Township in Delaware County</td></tr>
<tr><td colspan="2" align="center">Coordinates: 40°11′28″N 85°16′21″W</td></tr>
<tr><td>Country</td><td>United States</td></tr>
<tr><td>State</td><td>Indiana</td></tr>
<tr><td>County</td><td>Delaware</td></tr>
<tr><td colspan="2">Government</td></tr>
<tr><td>• Type</td><td>Indiana township</td></tr>
<tr><td colspan="2">Area</td></tr>
<tr><td>• Total</td><td>35 sq mi (90.65 km^2)</td></tr>
<tr><td>• Land</td><td>34.91 sq mi (90.4 km^2)</td></tr>
<tr><td>• Water</td><td>0.09 sq mi (0.24 km^2)</td></tr>
<tr><td>Elevation[1]</td><td>1014 ft (309 m)</td></tr>
<tr><td colspan="2">Population (2000)</td></tr>
<tr><td>• Total</td><td>4,919</td></tr>
<tr><td>• Density</td><td>140.93/sq mi (54.41/km^2)</td></tr>
<tr><td>FIPS code</td><td>18-43236[2]</td></tr>
<tr><td>GNIS feature ID</td><td>453552</td></tr>
</table>

Liberty Township is one of twelve townships in Delaware County, Indiana. As of the 2000 census, its population was 4,919.

Geography

Liberty Township covers an area of 35 square miles (91 km^2); 0.09 square miles (0.23 km^2) (0.26 percent) of this is water.

Cities and towns

- Muncie (east edge)
- Selma

Unincorporated towns

- Hyde Park
- Smithfield
- Woodland Park

Adjacent townships

- Delaware Township (north)
- Monroe Township, Randolph County (east)
- Stoney Creek Township, Randolph County (southeast)
- Perry Township (south)
- Monroe Township (southwest)
- Center Township (west)
- Hamilton Township (northwest)

Major highways

- **32** Indiana State Road 32

Cemeteries

The township contains eight cemeteries: Bortsfield, Freidline, Graham, Mount Tabor, Orr, Sparr, Truitt and White.

References

- "Liberty Township, Delaware County, Indiana" [3]. Geographic Names Information System, U.S. Geological Survey. Retrieved 2009-09-24.
- United States Census Bureau cartographic boundary files [4]

[1] "US Board on Geographic Names" (http://geonames.usgs.gov). United States Geological Survey. 2007-10-25. . Retrieved 2008-01-31.

[2] "American FactFinder" (http://factfinder.census.gov). United States Census Bureau. . Retrieved 2008-01-31.

[3] http://geonames.usgs.gov/pls/gnispublic/f?p=gnispq:3:::NO::P3_FID:453552

[4] http://www.census.gov/geo/www/cob/

External links

- Indiana Township Association (http://www.indianatownshipassoc.org/)
- United Township Association of Indiana (http://unitedtownships.org/)

Delaware County, Indiana

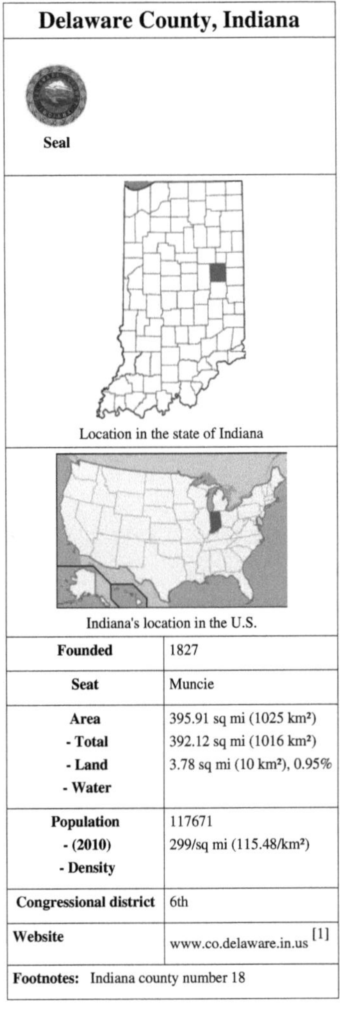

Delaware County, Indiana

Seal

Location in the state of Indiana

Indiana's location in the U.S.

Founded	1827
Seat	Muncie
Area - Total - Land - Water	395.91 sq mi (1025 km²) 392.12 sq mi (1016 km²) 3.78 sq mi (10 km²), 0.95%
Population - (2010) - Density	117671 299/sq mi (115.48/km²)
Congressional district	6th
Website	www.co.delaware.in.us [1]
Footnotes:	Indiana county number 18

Delaware County is a county located in the U.S. state of Indiana. As of 2010, the population was 117,671.[2] The county seat is Muncie.[3] It is part of the Muncie, IN, Metropolitan Statistical Area.

History

Muncie from the northwest.

Delaware County was formed in 1827. It was named for the Delaware, a Native American people who still lived in the county. The Delaware were removed from the county in the 1840s. The county was once home to The Prophet, the brother of Tecumseh who instigated a native uprising in 1811. David Conner was the first settler to live in the county in the early 1810s.[4]

Following the American Civil War the county experienced an economic boom after the discovery of natural gas that fueled rapid industrial growth in the surrounding area.

The first discovery of natural gas in Indiana occurred in the town of Eaton, in 1876. A company was drilling for coal and when they had reaching a depth of six-hundred feet, there was a great noise and bad smelling fumes began to come from the hole. After a partial investigation, many concluded that they had breached the ceiling of Hell, and the hole was quickly filled in. In 1884, when natural gas was discovered in nearby Ohio, the townsfolk recalled the incident and returned to the location and opened the state of Indiana's first natural gas well. The gas was so abundant and strong, that when the well was lit, the flames could be see from Muncie.[5]

Geography

According to the 2010 census, the county has a total area of 395.91 square miles (1025.4 km^2), of which 392.12 square miles (1015.6 km^2) (or 99.04%) is land and 3.78 square miles (9.8 km^2) (or 0.95%) is water.[6]

Cities and towns

- Albany
- Daleville
- Eaton
- Gaston
- Muncie
- Selma
- Yorktown

Townships

- Center
- Delaware
- Hamilton
- Harrison
- Liberty
- Monroe
- Mount Pleasant
- Niles
- Perry
- Salem
- Union
- Washington

Major highways

- Interstate 69
- U.S. Route 35
- Indiana State Road 3
- Indiana State Road 28
- Indiana State Road 32
- Indiana State Road 67
- Indiana State Road 167
- Indiana State Road 332

Adjacent counties

- Blackford County (north)
- Jay County (northeast)
- Randolph County (east)
- Henry County (south)
- Madison County (west)
- Grant County (northwest)

Climate and weather

In recent years, average temperatures in Muncie have ranged from a low of 16 °F (−9 °C) in January to a high of 85 °F (29 °C) in July, although a record low of −29 °F (−34 °C) was recorded in January 1994 and a record high of 102 °F (39 °C) was recorded in June 1988. Average monthly precipitation ranged from 2.06 inches (52 mm) in January to 4.28 inches (109 mm) in June.[]

Government

The county government is a constitutional body, and is granted specific powers by the Constitution of Indiana, and by the Indiana Code.

County Council: The county council is the legislative branch of the county government and controls all the spending and revenue collection in the county. Representatives are elected from county districts. The council members serve four year terms. They are responsible for setting salaries, the annual budget, and special spending. The council also has limited authority to impose local taxes, in the form of an income and property tax that is subject to state level approval, excise taxes, and service taxes.[7] [8]

Board of Commissioners: The executive body of the county is made of a board of commissioners. The commissioners are elected county-wide, in staggered terms, and each serves a four-year term. One of the commissioners, typically the most senior, serves as president. The commissioners are charged with executing the acts legislated by the council, collecting revenue, and managing the day-to-day functions of the county government.[7] [8]

Court: The county maintains a small claims court that can handle some civil cases. The judge on the court is elected to a term of four years and must be a member of the Indiana Bar Association. The judge is assisted by a constable who is also elected to a four-year term. In some cases, court decisions can be appealed to the state level circuit court.[8]

County Officials: The county has several other elected offices, including sheriff, coroner, auditor, treasurer, recorder, surveyor, and circuit court clerk. Each of these elected officers serves a term of four years and oversees a different part of county government. Members elected to county government positions are required to declare party affiliations and to be residents of the county.[8]

Delaware County is part of Indiana's 6th congressional district; Indiana Senate district 26;[9] and Indiana House of Representatives districts 33, 34 and 35.[10]

Demographics

Historical populations			
Census	**Pop.**		**%±**
1820	3677		—
1830	2374		−35.4%
1840	8843		272.5%
1850	10843		22.6%
1860	15753		45.3%
1870	19030		20.8%
1880	22926		20.5%
1890	30131		31.4%
1900	49624		64.7%
1910	51414		3.6%
1920	56377		9.7%
1930	67270		19.3%
1940	74963		11.4%
1950	90252		20.4%
1960	110938		22.9%
1970	129219		16.5%
1980	128587		−0.5%
1990	119659		−6.9%
2000	118769		−0.7%
2010	117671		−0.9%
Sources: United States Department of Commerce, Bureau of the Census, Population Division[11] Census Quickfacts[2]			

As of the census[12] of 2000, there were 118,769 people, 47,131 households, and 29,692 families residing in the county. The population density was 302 people per square mile (117/km²). There were 51,032 housing units at an average density of 130 per square mile (50/km²). The racial makeup of the county was 90.73% White, 6.72% Black or African American, 0.23% Native American, 0.66% Asian, 0.05% Pacific Islander, 0.47% from other races, and 1.13% from two or more races. 1.10% of the population were Hispanic or Latino of any race. 22.7% were of American, 20.9% German, 12.2% English and 9.8% Irish ancestry according to Census 2000.

There were 47,131 households out of which 27.80% had children under the age of 18 living with them, 48.50% were married couples living together, 10.90% had a female householder with no husband present, and 37.00% were non-families. 28.20% of all households were made up of individuals and 10.40% had someone living alone who was 65 years of age or older. The average household size was 2.37 and the average family size was 2.90.

In the county the population was spread out with 22.10% under the age of 18, 16.90% from 18 to 24, 25.60% from 25 to 44, 21.90% from 45 to 64, and 13.50% who were 65 years of age or older. The median age was 34 years. For every 100 females there were 92.20 males. For every 100 females age 18 and over, there were 89.20 males.

The median income for a household in the county was $34,659, and the median income for a family was $45,394. Males had a median income of $36,155 versus $23,268 for females. The per capita income for the county was $19,233. About 9.00% of families and 15.10% of the population were below the poverty line, including 15.70% of those under age 18 and 8.00% of those age 65 or over.

See also

- National Register of Historic Places listings in Delaware County, Indiana

References

[1] http://www.co.delaware.in.us
[2] "Delaware County QuickFacts" (http://quickfacts.census.gov/qfd/states/18/18035.html). United States Census Bureau. . Retrieved 2011-09-17.
[3] "Find a County" (http://www.naco.org/Counties/Pages/FindACounty.aspx). National Association of Counties. . Retrieved 2011-06-07.
[4] De Witt Clinton Goodrich & Charles Richard Tuttle (1875). *An Illustrated History of the State of Indiana* (http://books.google.com/books?id=YDIUAAAAYAAJ). Indiana: R. S. Peale & co.. pp. 556. .
[5] Gray, Ralph (1995). *Indiana History: A Book of Readings*. Indiana University Press. pp. 187. ISBN 025332629X.
[6] "Census 2010 U.S. Gazetteer Files: Counties" (http://www.census.gov/geo/www/gazetteer/files/Gaz_counties_national.txt). United States Census. . Retrieved 2011-10-10.
[7] Indiana Code. "Title 36, Article 2, Section 3" (http://www.in.gov/legislative/ic/code/title36/ar2/ch3.html). IN.gov. . Retrieved 2008-09-16.
[8] Indiana Code. "Title 2, Article 10, Section 2" (http://www.in.gov/legislative/ic/code/title3/ar10/ch2.pdf). IN.gov. . Retrieved 2008-09-16.
[9] "Indiana Senate Districts" (http://www.in.gov/sos/elections/3006.htm). State of Indiana. . Retrieved 2011-01-23.
[10] "Indiana House Districts" (http://www.in.gov/sos/elections/3005.htm). State of Indiana. . Retrieved 2011-01-23.
[11] Forstall, Richard L. (editor) (1996). *Population of states and counties of the United States: 1790 to 1990 : from the twenty-one decennial censuses* (http://books.google.com/books?id=Z12v1lrkv2IC&lpg=PA50&pg=PA50#v=onepage&q&f=false). United States Department of Commerce, Bureau of the Census, Population Division. pp. 50–53. ISBN 0-934213-48-8. .
[12] "American FactFinder" (http://factfinder.census.gov). United States Census Bureau. . Retrieved 2008-01-31.

External links

- Delaware County, Indiana Geographic Information System (http://www.co.delaware.in.us/department/?fDD=25-0)
- Delaware County Weather (http://www.muncieweather.com)
- Muncie Free Press (http://www.munciefreepress.com) - Delaware County News and Information
- Downtown Muncie (http://www.munciedowntown.com/)
- ScanMuncie - Online Community Forums & Online Public Safety Radio Scanners (http://scanmuncie.com/)
- Ball State University - External link (http://www.bsu.edu/)
- Ball State University Libraries - External link (http://www.bsu.edu/library/)

Muncie, Indiana

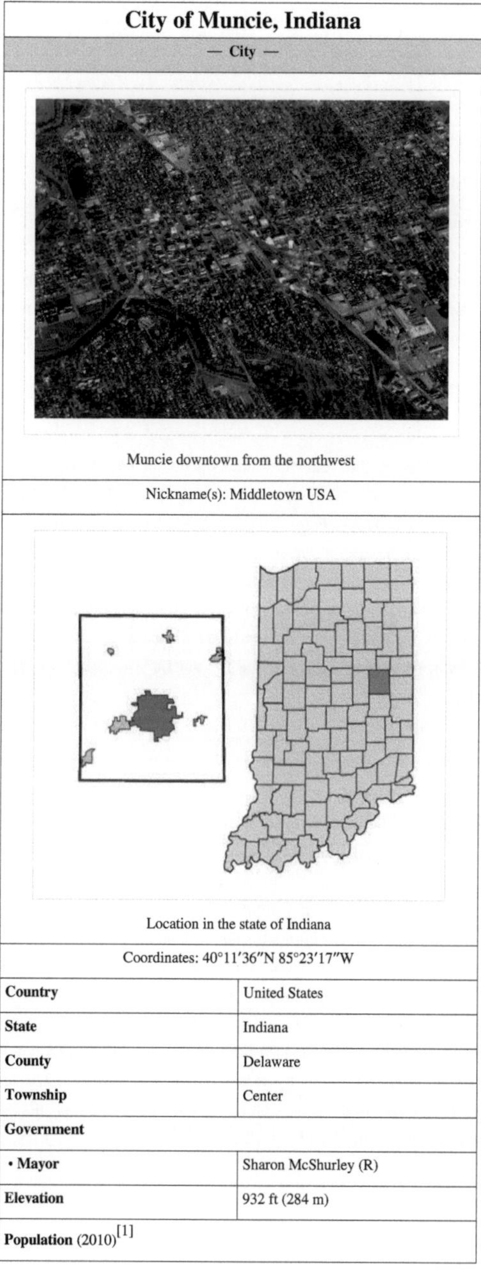

City of Muncie, Indiana	
— City —	

Muncie downtown from the northwest

Nickname(s): Middletown USA

Location in the state of Indiana

Coordinates: 40°11′36″N 85°23′17″W

Country	United States
State	Indiana
County	Delaware
Township	Center
Government	
• **Mayor**	Sharon McShurley (R)
Elevation	932 ft (284 m)
Population (2010)[1]	

• Total	70,085
• Demonym	Munsonian
Time zone	EST (UTC-5)
• Summer (DST)	EDT (UTC-4)
ZIP codes	47302-47308
Area code(s)	765
FIPS code	18-51876[2]
GNIS feature ID	0439878[3]
Website	www.cityofmuncie.com [4]

Muncie (◀ /ˈmʌnsi/) is a city in Center Township, Delaware County in east central Indiana, best known as the home of Ball State University and the birthplace of the Ball Corporation. It is the principal city of the Muncie, Indiana, Metropolitan Statistical Area, which has a population of 118,769. The city population, as of the 2010 Census, is 70,085.[1]

History

The area was first settled in the 1770s by the Delaware Indians, who had been transported from their tribal lands near the east coast to Ohio and eastern Indiana. They founded several towns along the White River including Munsee Town (according to historical map of "The Indians" by Clark Ray), near the site of present-day Muncie. The tribes were forced to cede their land to the federal government and move farther west in 1818, and in 1820 the area was opened to white settlers. Muncie was one of the considerations for state capital when it was moved from Corydon. It was considered by many to be a suitable location due to its location on the White River. The city of Muncie was incorporated in 1865. Contrary to popular legend, the city is not named after a mythological Chief Munsee, rather it was named after Munsee Town, the white settlers' name for the Indian village on the site, "munsee" meaning a member of the Delaware tribe.

*Note: **Munsee** is one of the Algonquian languages (nearly extinct) spoke by the Lenape (Delaware).*

Middletown studies

Muncie was lightly disguised as "Middletown" by a team of sociologists, led by Robert and Helen Lynd, who were only the first to conduct a series of studies in Muncie—considered a typical Middle-American community—in their case, a study funded by the Rockefeller Institute of Social and Religious Research.[5] In 1929, the Lynds published *Middletown: A Study in Contemporary American Culture.* They returned to re-observe the community during the Depression and published *Middletown in Transition: A Study in Cultural Conflicts* (1937). Later in the century, the National Science Foundation funded a third major study that resulted in two books by Theodore Caplow, *Middletown Families* (1982) and *All Faithful People* (1983). Caplow returned in 1998 to begin another study, Middletown IV, which became part of a PBS Documentary entitled "The First Measured Century," released in December 2000. The Ball State Center for Middletown Studies [6] continues to survey and analyze social change in Muncie. An enormous database of the Middletown surveys conducted between 1978 and 1997 is available online from ARDA, American Religion Data Archive [7]. Ironically, a Henry County farming community actually called Middletown is only a 20-minute drive from Muncie.

Demographics

Historical populations		
Census	Pop.	%±
1850	606	—
1860	1782	194.1%
1870	2992	67.9%
1880	5219	74.4%
1890	11345	117.4%
1900	20942	84.6%
1910	24005	14.6%
1920	36524	52.2%
1930	46548	27.4%
1940	49720	6.8%
1950	58479	17.6%
1960	68603	17.3%
1970	69082	0.7%
1980	76460	10.7%
1990	71035	−7.1%
2000	67430	−5.1%
2010	70085	3.9%

As of the 2010 United States Census, the population is 70,085.

As of the 2000 census, the population was 67,430. There were 27,322 households, and 14,589 families residing in the city. The population density was 2,788.2 people per square mile (1,076.7/km²). There were 30,205 housing units at an average density of 1,248.9 per square mile (482.3/km²). The racial makeup of the city was 83.72% White, 12.97% African American, 0.27% Native American, 0.79% Asian, 0.09% Pacific Islander, 0.67% from other races, and 1.49% from two or more races. Hispanic or Latino of any race were 1.44% of the population.

There were 27,322 households out of which 23.7% had children under the age of 18 living with them, 36.4% were married couples living together, 13.0% had a female householder with no husband present, and 46.6% were non-families. 34.1% of all households were made up of individuals and 11.8% had someone living alone who was 65 years of age or older. The average household size was 2.24 and the average family size was 2.86.

In the city the population was spread out with 19.8% under the age of 18, 24.6% from 18 to 24, 24.2% from 25 to 44, 18.3% from 45 to 64, and 13.2% who were 65 years of age or older. The median age was 29 years. For every 100 females there were 89.9 males. For every 100 females age 18 and over, there were 86.5 males.

The median income for a household in the city was $26,613, and the median income for a family was $36,398. Males had a median income of $30,445 versus $21,872 for females. The per capita income for the city was $15,814. About 14.3% of families and 23.1% of the population were below the poverty line, including 24.2% of those under age 18 and 9.7% of those age 65 or over.

Federally, Muncie is part of Indiana's 6th congressional district, represented by Republican Mike Pence, elected in 2000.

The state's senior member of the United States Senate is Republican Richard Lugar, elected in 1976. The state's junior member of the United States Senate is Republican Dan Coats, elected in 2010. The Governor of Indiana is Republican Mitch Daniels, elected in 2004.

Economy

Since the late 19th century, Muncie's economic backbone had been the in the industrial sector, primarily in manufacturing. Drawn to the region during the Indiana Gas Boom of the 1880s, many factories sprang up in the area that relied on the combustible natural resource. The Ball Brothers moved their glass factory from Buffalo to Muncie, beginning glass production there on March 1, 1888.[8] This relationship with Muncie ended 110 years later, as Ball Corporation moved its corporation headquarters to Broomfield, Colorado in 1998. Other notable factories that were located in Muncie include: Delco Remy, Westinghouse (later ABB), Indiana Steel and Wire, General Motors (New Venture Gear), Warner Gear (later BorgWarner), Broderick Co. Inc., Dayton-Walter, and Ball Corporation. However, most of these factories closed during a tumultuous period for the city from the late 1980s and late 1990s. As of 2010, none of the aforementioned factories operated within Muncie. However, many smaller, non-unionized, manufacturing businesses have survived this transition such as Maxon Corporation, Duffy Tool (now North American Stamping), Reber Machine & Tool, MAGNA Powertrain, and a dozen or so other shops that employ anywhere from a few dozen to a couple of hundred workers.

Like many mid-sized cities in the Rust Belt, Muncie has had to economically reinvent itself due to the collective fall of the manufacturing industry in the latter part of the 20th century. Muncie's current economic backbone is in health care, education, retail, and other service industries. The largest employers in Muncie are Ball Memorial Hospital (an Indiana University Health partner), Ball State University, Muncie Community Schools, The City of Muncie, Sallie Mae, Wal-mart, and The Youth Opportunity Center. In 2008, Italian manufacturer Brevini Power Transmission announced that Muncie will be its new U.S. headquarters and plans to create 450 jobs in Muncie by 2011.[9]

Ball Memorial Hospital Complex

The local economy is one of the most controversial topics for Muncie residents, and the city has at times struggled to find cohesion between older unemployed/underemployed Muncie residents who strongly identify with the manufacturing-identity of the city, and newer residents who identify with the city's shift towards educational and health services. Animosity is greatest amongst those in the older, once industrialized parts on the south and east parts of town as much of the economic growth over that last 20+ years has taken place primarily on the northwest portions of town in connection with the growth of both Ball Memorial Hospital and Ball State University. Muncie, once a factory town with a small teacher's college, is now considered by many as a college-town with a manufacturing past.

Sports

Muncie has gained notoriety as having a rich tradition in prep sports. Muncie Central High School has fielded a boys basketball team for over 100 years and is the most successful such program, with more state championships (8 State Titles, 7 runner-ups) in the state noted for boys' high school basketball and Hoosier Hysteria.[10] The "Bearcats" of Muncie Central High School has called the Walnut St. Fieldhouse home since 1928. The 6,000+ (once 7,600) seat facility was one of the largest facilities of its kind when built, and still ranks in the top 20 in being the largest high school gymnasium in the world.[11] Muncie Central also boasts 6 state championships in girls volleyball. Burris Laboratory School has also gained notoriety on a national level for its girls volleyball program. The elite program has won 21 state championships, including the last 13 2A state titles, as well as 4 national championships all under the helm of former coach Steve Shondell.[12] Muncie Southside High School also has had success in winning two Wrestling State Championships (1975 & 1990) as well as a runner-up finish in the Class 3A Boys Basketball State Championship (2001). Lost to consolidation in 1988, Muncie Northside High School also found success in athletics winning three Girls' Volleyball State Championships (1975–1978) and one Wrestling State Championship (1974). [13]

Professionally, Muncie was once home to a National Football League team. The Muncie Flyers (also known as the Congerville Flyers) were professional football team from 1905–1925 and were one of the 11 charter members of the NFL, playing in the league from 1920-1924.[14]

Education

Elementary schools

- Burris Laboratory School
- East Washington Academy
- South View Elementary
- Grissom Elementary
- Storer Elementary
- Longfellow Elementary
- Sutton Elementary
- Mitchell Elementary
- North View Elementary
- West View Elementary
- Heritage Hall Christian School
- Hoosier Academy Muncie
- St. Lawrence Elementary School
- St. Mary Elementary School

Shafer Tower on the campus of Ball State.

Middle schools

- Burris Laboratory School
- Northside Middle School
- Wilson Middle School
- Heritage Hall Christian School
- Hoosier Academy Muncie
- Pope John Paul II Middle School

High schools

- Burris Laboratory School
- Heritage Hall Christian School
- Indiana Academy for Science, Mathematics, and Humanities
- Muncie Central High School
- Muncie Southside High School

Muncie City Hall, 2005.

For other Delaware County high schools, click here.

Colleges and universities

- Ball State University
- Ivy Tech Community College
- Harrison College (Indiana)

Notable natives & residents

The former C&O depot, restored and now used as the office for the adjacent bicycle trail.

General

- Ball Brothers, founders of the Ball Corporation, originally a producer of glass canning jars but now a producer of various products.
- Benjamin V. Cohen - a key figure in the administrations of Presidents Franklin D. Roosevelt and Harry S. Truman.
- George R. Dale, editor of the Muncie *Post-Democrat* (1920–1936) who gained national attention by speaking out against the Ku Klux Klan.[15]
- Bertha Fry - At the time of her death on November 14, 2007, the 3rd oldest person living on earth at 113 years.[16]

Arts

- Ray Boltz - Contemporary Christian musical artist [17]
- Angelin Chang, GRAMMY®-award winning classical pianist
- Trevor Chowning - Pop artist and former Hollywood talent agent/producer
- Jim Davis - cartoonist, creator of the Garfield comic strip, which has become popular since its debut in June 1978.[18] Attended Ball State.[19]
- Emily Kimbrough - Noted author and magazine editor. Author of *Our Hearts Were Young and Gay* and *How Dear to My Heart* a recount of her childhood in Muncie.[20]

Sports

- Ron Bonham -- Former All-American Muncie Central HS basketball standout. Played on NCAA champion Cincinnati Bearcats and NBA Champion Boston Celtics teams in 1960s.
- Bill Dinwiddie - National Basketball Association forward
- Dave Duerson - All-American Defensive Back for the University of Notre Dame; played 11 seasons in the NFL with the Chicago Bears, New York Giants and Phoenix Cardinals.[21]
- Brandon Gorin - National Football League offensive lineman [22]
- Adam Lind - professional baseball player for the Toronto Blue Jays
- Matt Painter - Purdue men's basketball head coach[23]
- Bonzi Wells - Former Muncie Central High School and Ball State University standout. Drafted 11th overall in the 1998 NBA draft. Currently plays for the Puerto Rican team Capitanes de Arecibo.[1] [24]
- Ryan Kerrigan - Former Muncie Central High School and Purdue University football standout. Drafted 16th by the Washington Redskins in the 2011 NFL Draft.

Popular culture

Figures from Muncie have appeared on several American movies, such as:

* Everything You Always Wanted to Know About Sex* (*But Were Afraid to Ask) - On the show *What's My Perversion?*, contestant Chaim Baumel is from Muncie.

* In the 1994 film *The Hudsucker Proxy*, the main protagonist Norville Barnes (Tim Robbins) grew up in Muncie and graduated from the fictional Muncie College of Business Administration. Coincidentally, reporter Amy Archer (Jennifer Jason Leigh) wrongly claims to be from there. As a result, she has to sing with Barnes the song of the local college sports team, the Eagles.

See also

* Cincinnati, Richmond & Muncie Depot
* Hemingray Glass Company
* List of public art in Muncie, Indiana
* Muncie Indiana Transit System
* WWHI (WCRD), Ball State University's Radio Station, 91.3 MHz FM
* WIPB, A Public Broadcasting Station owned by Ball State University
* WMUN-LP, A low power broadcast television station affiliated with TBN in Muncie
* WLBC, A Muncie Radio station 104.1 MHz FM
* WERK, A Muncie Oldies station 104.9 MHz FM

References

[1] "U.S. Census Bureau Delivers Indiana's 2010 Census Population Totals" (http://2010.census.gov/news/releases/operations/cb11-cn26.html). . Retrieved 11 February 2011.

[2] "American FactFinder" (http://factfinder.census.gov). United States Census Bureau. . Retrieved 2008-01-31.

[3] "US Board on Geographic Names" (http://geonames.usgs.gov). United States Geological Survey. 2007-10-25. . Retrieved 2008-01-31.

[4] http://www.cityofmuncie.com/

[5] "The aim... was to study synchronously the interwoven trends that are the life of a small American city." Lynd and Lynd 1929: 3

[6] http://www.bsu.edu/middletown/

[7] http://www.thearda.com/

[8] Hoover, Dwight W., A pictorial history of Indiana, Indiana University Press, 1980

[9] Trade and Industry Development, http://www.tradeandindustrydev.com/ID-915-news.aspx

[10] Stodghill, Dick and Jackie, BEARCATS!: A History of Basketball at Muncie Central High School, JLT Publications, 1988

[11] USA Today, http://www.usatoday.com/sports/preps/basketball/2004-02-25-ten-great-hoops-newcastle_x.htm

[12] The Indianapolis Star,http://www.usatoday.com/sports/preps/basketball/2004-02-25-ten-great-hoops-newcastle_x.htm

[13] IHSAA,http://www.ihsaa.org/dnn/Schools/StateChampHistory/StateChampionshipsbySchool/tabid/581/Default.aspx

[14] The Muncie Flyers,http://krd4052.freewebhostx.com/MuncieFlyers.htm

[15] Ball State University Archives (http://www.bsu.edu/libraries/viewpage.aspx?src=./collections/archives/george_dale.html)

[16] The article requested can not be found! Please refresh your browser or go back. (C7,20080325,,80214016,AR). | The Star Press - www.thestarpress.com - Muncie, IN (http://www.thestarpress.com/apps/pbcs.dll/article?AID=200771114029)

[17] Ray Boltz (http://www.rayboltz.com)

[18] The Official Website of Garfield and Friends (http://www.garfield.com/about/jim.html)

[19] Jim Davis :: Profile (http://www.bsu.edu/cob/profile/0,1391,5289-389-64102,00.html)

[20] http://query.nytimes.comgst/fullpage.html?res=950DE7DC123EF932A25751C0A96F948260 Emily Kimbrough

[21] Dave Duerson Past Stats, Statistics, History, and Awards - databaseFootball.com (http://databasefootball.com/players/playerpage.htm?ilkid=DUERSDAV01)

[22] Brandon Gorin | NFL Football at CBSSports.com (http://cbs.sportsline.com/nfl/players/playerpage/235067)

[23] Player Bio: Matt Painter :: Men's Basketball (http://purduesports.cstv.com/sports/m-baskbl/mtt/painter_matt00.html)

[24] Bonzi Wells Statistics - Basketball-Reference.com (http://www.basketball-reference.com/players/w/wellsbo01.html)

External links

- City of Muncie, Indiana website (http://www.cityofmuncie.com/)
- Muncie Chamber of Commerce (http://www.muncie.com/)
- The Star Press (http://www.thestarpress.com/)
- Downtown Muncie Website (http://www.munciedowntown.us/)
- Muncie Weather Website (http://www.muncieweather.com/)
- The Muncie Scene (http://www.themunciescene.com/), art/music community portal
- "The Lynds Revisited" by Richard Jensen, in Indiana Magazine of History (Dec 1979) 75: 303-319 (http://members.aol.com/dann01/lynds.html)
- Delaware County Office of Geographic Information (http://www.co.delaware.in.us/DCGIS/index.html)
- The Emily Kimbrough Historic District (http://www.muncie-ecna.org/ecna/kimbrough.php)
- Muncie, Indiana travel guide from Wikitravel
- Chisholm, Hugh, ed (1911). "Muncie". *Encyclopædia Britannica* (11th ed.). Cambridge University Press.
- LIFE Magazine May 10, 1937 (http://books.google.com/books?id=vEQEAAAAMBAJ&printsec=frontcover&source=gbs_ge_summary_r&cad=0#v=onepage&q&f=false), "Middletown-Muncie", pages 15–25, ("the Picture Essay"), at Google Books.
- Digitized archival collections related to Muncie and its history (http://libx.bsu.edu/cdm4/BrowseResults.php?sortby=SUBJECT#Middletown Studies) available in the Ball State University Digital Media Repository

Albany, Indiana

Town of Albany, Indiana

— Town —

Albany from the west, with Redkey in the background.

Location in the state of Indiana

Coordinates: 40°18′5″N 85°14′16″W

Country	United States
State	Indiana
Counties	Delaware, Randolph
Townships	Niles, Delaware
Area	
• **Total**	1.7 sq mi (4.3 km^2)
• **Land**	1.7 sq mi (4.3 km^2)
• **Water**	0.0 sq mi (0.0 km^2)
Elevation	909 ft (277 m)

Population (2010)	
• Total	2165
• Density	1435.1/sq mi (554.1/km^2)
Time zone	EST (UTC−5)
• Summer (DST)	EST (UTC−5)
ZIP code	47320
Area code(s)	765
FIPS code	18-00802[1]
GNIS feature ID	0430041[2]

Albany is a town in Delaware and Randolph counties in the U.S. state of Indiana, along the Mississinewa River. The population was 2,165 at the 2010 census. It is part of the Muncie, IN Metropolitan Statistical Area.

Geography

Albany is located at 40°18′5″N 85°14′16″W (40.30, -85.24)[3] .

According to the United States Census Bureau, the town has a total area of 1.6 square miles (4.1 km^2), all of it land.

Demographics

As of the census[1] of 2000, there were 2,368 people, 958 households, and 646 families residing in the town. The population density was 1,434.2 people per square mile (554.1/km²). There were 1,038 housing units at an average density of 628.7 per square mile (242.9/km²). The racial makeup of the town was 98.44% White, 0.42% African American, 0.08% Native American, 0.04% Asian, 0.08% from other races, and 0.93% from two or more races. Hispanic or Latino of any race were 0.13% of the population.

There were 958 households out of which 30.8% had children under the age of 18 living with them, 53.8% were married couples living together, 10.3% had a female householder with no husband present, and 32.5% were non-families. 28.8% of all households were made up of individuals and 13.7% had someone living alone who was 65 years of age or older. The average household size was 2.39 and the average family size was 2.95.

In the town the population was spread out with 24.4% under the age of 18, 7.9% from 18 to 24, 28.2% from 25 to 44, 22.3% from 45 to 64, and 17.3% who were 65 years of age or older. The median age was 38 years. For every 100 females there were 83.4 males. For every 100 females age 18 and over, there were 83.1 males.

The median income for a household in the town was $33,314, and the median income for a family was $40,893. Males had a median income of $33,929 versus $24,286 for females. The per capita income for the town was $16,620. About 4.5% of families and 5.9% of the population were below the poverty line, including 7.9% of those under age 18 and 1.3% of those age 65 or over.

Car shows

Each summer Albany plays host to a series of "cruise-in" car show events at the McDonald's on the west side of the city. The events, hosted by the Crown City Cruisers car club from nearby Dunkirk, draw cars from all over the state of Indiana and many from nearby Ohio.

References

[1] "American FactFinder" (http://factfinder.census.gov). United States Census Bureau. . Retrieved 2008-01-31.

[2] "US Board on Geographic Names" (http://geonames.usgs.gov). United States Geological Survey. 2007-10-25. . Retrieved 2008-01-31.

[3] "US Gazetteer files: 2010, 2000, and 1990" (http://www.census.gov/geo/www/gazetteer/gazette.html). United States Census Bureau. 2011-02-12. . Retrieved 2011-04-23.

External links

- Crown City Cruisers car club (http://www.crowncitycruisers.org/)

Chesterfield, Indiana

Chesterfield

— Town —

Town of Chesterfield

Aerial view of Chesterfield facing southwest

Location in the state of Indiana

Coordinates: 40°6′44″N 85°35′47″W

Country	United States
State	Indiana
Counties	Madison, Delaware
Townships	Union,
Area	
• Total	1.1 sq mi (3.0 km^2)
• Land	1.1 sq mi (3.0 km^2)

• **Water**	0.0 sq mi (0.0 km^2)
Elevation	909 ft (277 m)
Population (2010)	
• **Total**	2547
• **Density**	2581.7/sq mi (996.8/km^2)
Time zone	EST (UTC-5)
• **Summer (DST)**	EST (UTC-5)
ZIP code	46017
Area code(s)	765
FIPS code	18-12376[1]
GNIS feature ID	0432454[2]
Website	http://chesterfield.in.gov

Chesterfield is a town in the U.S. state of Indiana which lies in Union Township, Madison County. The population was 2,547 at the 2010 census. It is part of the Anderson, Indiana Metropolitan Statistical Area.

Geography

Chesterfield is located at 40°6′44″N 85°35′47″W (40.112193, -85.596324)[3].

According to the United States Census Bureau, the town has a total area of 1.1 square miles (2.8 km^2), all of it land.

Demographics

As of the census[1] of 2000, there were 2,969 people, 1,269 households, and 808 families residing in the town. The population density was 2,586.5 people per square mile (996.8/km²). There were 1,365 housing units at an average density of 1,189.1 per square mile (458.3/km²). The racial makeup of the town was 98.42% White, 0.40% African American, 0.10% Native American, 0.30% Asian, 0.30% from other races, and 0.47% from two or more races. Hispanic or Latino of any race were 1.38% of the population.

There were 1,269 households out of which 31.3% had children under the age of 18 living with them, 48.3% were married couples living together, 11.4% had a female householder with no husband present, and 36.3% were non-families. 30.8% of all households were made up of individuals and 11.6% had someone living alone who was 65 years of age or older. The average household size was 2.30 and the average family size was 2.87.

In the town the population was spread out with 24.2% under the age of 18, 8.5% from 18 to 24, 29.9% from 25 to 44, 22.5% from 45 to 64, and 14.9% who were 65 years of age or older. The median age was 36 years. For every 100 females there were 87.7 males. For every 100 females age 18 and over, there were 84.3 males.

The median income for a household in the town was $37,143, and the median income for a family was $47,222. Males had a median income of $35,412 versus $22,966 for females. The per capita income for the town was $18,738. About 2.9% of families and 5.9% of the population were below the poverty line, including 5.9% of those under age 18 and 7.5% of those age 65 or over.

Transport

Chesterfield is served by a variety of major roads.

- Interstate 69 (exit 34)
- 32 State Road 32
- 67 State Road 67
- 232 State Road 232

References

[1] "American FactFinder" (http://factfinder.census.gov). United States Census Bureau. . Retrieved 2008-01-31.

[2] "US Board on Geographic Names" (http://geonames.usgs.gov). United States Geological Survey. 2007-10-25. . Retrieved 2008-01-31.

[3] "US Gazetteer files: 2010, 2000, and 1990" (http://www.census.gov/geo/www/gazetteer/gazette.html). United States Census Bureau. 2011-02-12. . Retrieved 2011-04-23.

External links

- Town of Chesterfield, Indiana website (http://chesterfield.in.gov)
- Chesterfield-Union Township Fire Department (http://www.cutfd.org)

Daleville, Indiana

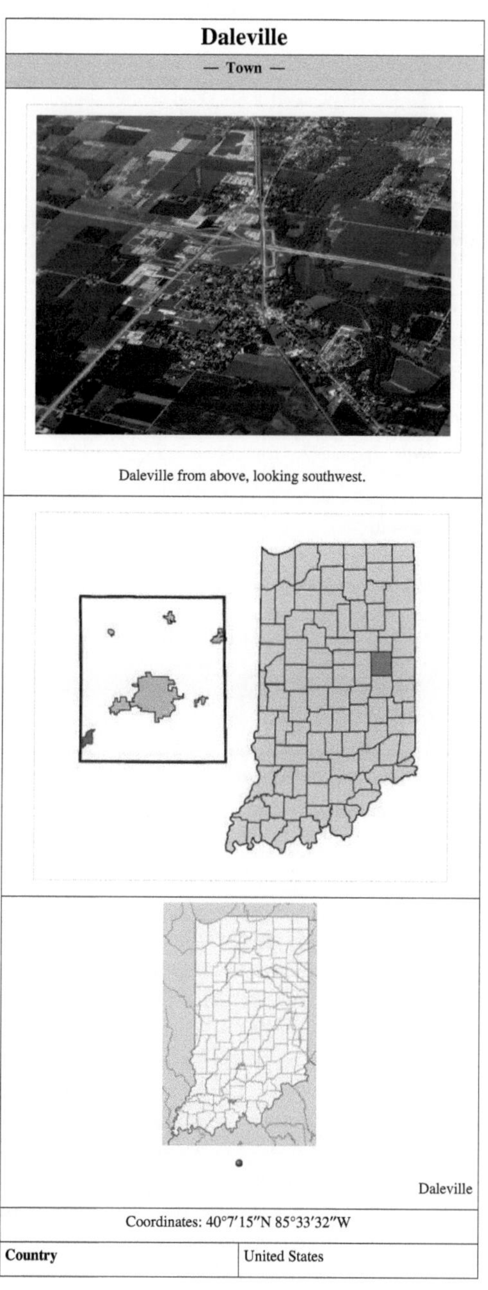

<table>
<tr><td colspan="2" align="center">Daleville</td></tr>
<tr><td colspan="2" align="center">— Town —</td></tr>
<tr><td colspan="2" align="center">Daleville from above, looking southwest.</td></tr>
<tr><td colspan="2" align="right">Daleville</td></tr>
<tr><td colspan="2" align="center">Coordinates: 40°7′15″N 85°33′32″W</td></tr>
<tr><td>Country</td><td>United States</td></tr>
</table>

State	Indiana
County	Delaware
Township	Salem
Area	
• Total	2.0 sq mi (5.2 km^2)
• Land	2.0 sq mi (5.2 km^2)
• Water	0.0 sq mi (0.1 km^2)
Elevation	912 ft (278 m)
Population (2010)	
• Total	1647
• Density	833.2/sq mi (321.7/km^2)
Time zone	EST (UTC-5)
• Summer (DST)	EST (UTC-5)
ZIP code	47334
Area code(s)	765
FIPS code	18-16642[1]
GNIS feature ID	0433295[2]

Daleville is a town in Salem Township, Delaware County, Indiana, United States. The population was 1,647 at the 2010 census. It is part of the Muncie, IN Metropolitan Statistical Area.

Geography

Daleville is located at 40°07′15″N 85°33′32″W.[3]

According to the United States Census Bureau, the town has a total area of 2.0 square miles (5.2 km^2), of which, 2.0 square miles (5.2 km^2) of it is land and 0.04 square miles (0.10 km^2) of it (1.00%) is water.

Daleville, Indiana is the newest town in Indiana, Daleville was not officially incorporated as a town until 1982

Demographics

As of the census[1] of 2000, there were 1,658 people, 650 households, and 453 families residing in the town. The population density was 832.4 people per square mile (321.7/km²). There were 688 housing units at an average density of 345.4 per square mile (133.5/km²). The racial makeup of the town was 98.97% White, 0.12% African American, 0.24% Native American, 0.06% Asian, 0.36% from other races, and 0.24% from two or more races. Hispanic or Latino of any race were 0.72% of the population.

There were 650 households out of which 35.2% had children under the age of 18 living with them, 57.2% were married couples living together, 9.5% had a female householder with no husband present, and 30.2% were non-families. 25.8% of all households were made up of individuals and 11.5% had someone living alone who was 65 years of age or older. The average household size was 2.55 and the average family size was 3.09.

In the town the population was spread out with 28.0% under the age of 18, 8.1% from 18 to 24, 30.3% from 25 to 44, 21.8% from 45 to 64, and 11.8% who were 65 years of age or older. The median age was 35 years. For every 100 females there were 90.6 males. For every 100 females age 18 and over, there were 89.7 males.

The median income for a household in the town was $40,592, and the median income for a family was $48,289. Males had a median income of $36,500 versus $23,182 for females. The per capita income for the town was $18,020. About 2.3% of families and 2.7% of the population were below the poverty line, including 1.6% of those under age 18 and 6.2% of those age 65 or over.

References

[1] "American FactFinder" (http://factfinder.census.gov). United States Census Bureau. . Retrieved 2008-01-31.

[2] "US Board on Geographic Names" (http://geonames.usgs.gov). United States Geological Survey. 2007-10-25. . Retrieved 2008-01-31.

[3] "US Gazetteer files: 2010, 2000, and 1990" (http://www.census.gov/geo/www/gazetteer/gazette.html). United States Census Bureau. 2011-02-12. . Retrieved 2011-04-23.

Celebrities- Commercial Bank robbed by John Dillinger for $3,500.00 on July 17, 1933

External links

- Daleville Community Schools (http://www.daleville.k12.in.us/)

Eaton, Indiana

<table>
<tr><td colspan="2" align="center">Town of Eaton, Indiana</td></tr>
<tr><td colspan="2" align="center">— Town —</td></tr>
<tr><td colspan="2" align="center">Eaton from the air, looking northeast.</td></tr>
<tr><td colspan="2" align="center">Location in the state of Indiana</td></tr>
<tr><td colspan="2" align="center">Coordinates: 40°20′23″N 85°21′13″W</td></tr>
<tr><td>Country</td><td>United States</td></tr>
<tr><td>State</td><td>Indiana</td></tr>
<tr><td>County</td><td>Delaware</td></tr>
<tr><td>Township</td><td>Union</td></tr>
<tr><td>Area</td><td></td></tr>
<tr><td>• Total</td><td>1.1 sq mi (2.9 km^2)</td></tr>
<tr><td>• Land</td><td>1.1 sq mi (2.9 km^2)</td></tr>
<tr><td>• Water</td><td>0.0 sq mi (0.0 km^2)</td></tr>
<tr><td>Elevation</td><td>886 ft (270 m)</td></tr>
</table>

Population (2010)	
• Total	1805
• Density	1431.2/sq mi (552.6/km^2)
Time zone	EST (UTC-5)
• Summer (DST)	EST (UTC-5)
ZIP code	47338
Area code(s)	765
FIPS code	18-20080[1]
GNIS feature ID	0433963[2]

Eaton is a town in Union Township, Delaware County, Indiana, along the Mississinewa River. The population was 1,805 at the 2010 census. It is part of the Muncie, IN Metropolitan Statistical Area.

History

The first discovery of natural gas in Indiana occurred in the town of Eaton in 1876. A company was drilling for coal and, when they had reaching a depth of six hundred feet, there was a great noise and bad-smelling fumes began to come from the hole. After a partial investigation, many concluded that they had breached the ceiling of Hell, and the hole was quickly filled in. In 1884, when natural gas was discovered in nearby Ohio, the townsfolk recalled the incident and returned to the location. There they opened the state of Indiana's first natural gas well. The gas was so abundant and strong that, when the well was lit, the flames could be seen from Muncie.[3] The discovery set off the Indiana Gas Boom, leading to two decades of rapid regional growth.

Geography

Eaton is located at 40°20′23″N 85°21′13″W (40.339675, -85.353746)[4] .

According to the United States Census Bureau, the town has a total area of 1.1 square miles (2.8 km^2), all of it land.

Demographics

As of the census[1] of 2000, there were 1,603 people, 619 households, and 459 families residing in the town. The population density was 1,432.0 people per square mile (552.6/km²). There were 661 housing units at an average density of 590.5 per square mile (227.9/km²). The racial makeup of the town was 98.75% White, 0.25% Native American, 0.06% Asian, 0.12% from other races, and 0.81% from two or more races. Hispanic or Latino of any race were 0.81% of the population.

There were 619 households out of which 40.2% had children under the age of 18 living with them, 57.2% were married couples living together, 11.0% had a female householder with no husband present, and 25.7% were non-families. 23.6% of all households were made up of individuals and 9.2% had someone living alone who was 65 years of age or older. The average household size was 2.59 and the average family size was 3.00.

In the town the population was spread out with 30.6% under the age of 18, 7.7% from 18 to 24, 29.8% from 25 to 44, 19.8% from 45 to 64, and 12.0% who were 65 years of age or older. The median age was 33 years. For every 100 females there were 96.7 males. For every 100 females age 18 and over, there were 91.4 males.

The median income for a household in the town was $31,563, and the median income for a family was $35,625. Males had a median income of $31,573 versus $20,645 for females. The per capita income for the town was $13,833. About 9.5% of families and 11.1% of the population were below the poverty line, including 13.9% of those under age 18 and 3.5% of those age 65 or over.

References

[1] "American FactFinder" (http://factfinder.census.gov). United States Census Bureau. . Retrieved 2008-01-31.

[2] "US Board on Geographic Names" (http://geonames.usgs.gov). United States Geological Survey. 2007-10-25. . Retrieved 2008-01-31.

[3] Gray, Ralph (1995). *Indiana History: A Book of Readings*. Indiana University Press. pp. 187. ISBN 025332629X.

[4] "US Gazetteer files: 2010, 2000, and 1990" (http://www.census.gov/geo/www/gazetteer/gazette.html). United States Census Bureau. 2011-02-12. . Retrieved 2011-04-23.

Gaston, Indiana

Town of Gaston, Indiana

— Town —

Gaston from the northeast.

Location in the state of Indiana

Coordinates: 40°18′49″N 85°30′3″W

Country	United States
State	Indiana
County	Delaware
Township	Washington
Area	
• Total	0.4 sq mi (0.9 km^2)
• Land	0.4 sq mi (0.9 km^2)
• Water	0.0 sq mi (0.0 km^2)
Elevation	886 ft (270 m)

Population (2010)	
• Total	871
• Density	2885.8/sq mi (1114.2/km^2)
Time zone	EST (UTC-5)
• Summer (DST)	EST (UTC-5)
ZIP code	47342
Area code(s)	765
FIPS code	18-27072[1]
GNIS feature ID	0434988[2]

Gaston is a town in Washington Township, Delaware County, Indiana, United States. The population was 871 at the 2010 census. It is part of the Muncie, IN Metropolitan Statistical Area.

Geography

Gaston is located at 40°18′49″N 85°30′3″W (40.313547, -85.500848)[3] .

According to the United States Census Bureau, the town has a total area of 0.3 square miles (0.78 km^2), all of it land.

Demographics

As of the census[1] of 2000, there were 1,010 people, 351 households, and 261 families residing in the town. The population density was 2,881.0 people per square mile (1,114.2/km²). There were 376 housing units at an average density of 1,072.5 per square mile (414.8/km²). The racial makeup of the town was 98.22% White, 0.59% African American, and 1.19% from two or more races. Hispanic or Latino of any race were 0.20% of the population.

There were 351 households out of which 44.4% had children under the age of 18 living with them, 51.3% were married couples living together, 17.4% had a female householder with no husband present, and 25.4% were non-families. 21.1% of all households were made up of individuals and 9.1% had someone living alone who was 65 years of age or older. The average household size was 2.67 and the average family size was 3.06.

In the town the population was spread out with 29.7% under the age of 18, 6.3% from 18 to 24, 31.6% from 25 to 44, 22.2% from 45 to 64, and 10.2% who were 65 years of age or older. The median age was 34 years. For every 100 females there were 102.4 males. For every 100 females age 18 and over, there were 100.6 males.

The median income for a household in the town was $31,853, and the median income for a family was $37,583. Males had a median income of $28,558 versus $23,281 for females. The per capita income for the town was $15,357. About 10.5% of families and 11.8% of the population were below the poverty line, including 18.3% of those under age 18 and 6.4% of those age 65 or over.

References

[1] "American FactFinder" (http://factfinder.census.gov). United States Census Bureau. . Retrieved 2008-01-31.

[2] "US Board on Geographic Names" (http://geonames.usgs.gov). United States Geological Survey. 2007-10-25. . Retrieved 2008-01-31.

[3] "US Gazetteer files: 2010, 2000, and 1990" (http://www.census.gov/geo/www/gazetteer/gazette.html). United States Census Bureau. 2011-02-12. . Retrieved 2011-04-23.

Yorktown, Indiana

<table>
<tr><td colspan="2" align="center">Yorktown, Indiana
— Town —</td></tr>
<tr><td colspan="2" align="center">Yorktown from the air, looking east.</td></tr>
<tr><td colspan="2" align="center">Location of Yorktown in the state of Indiana</td></tr>
<tr><td colspan="2" align="center">Coordinates: 40°10′31″N 85°28′55″W</td></tr>
<tr><td>Country</td><td>United States</td></tr>
<tr><td>State</td><td>Indiana</td></tr>
<tr><td>County</td><td>Delaware</td></tr>
<tr><td>Township</td><td>Mount Pleasant</td></tr>
<tr><td>Area</td><td></td></tr>
<tr><td>• Total</td><td>3.6 sq mi (9.3 km^2)</td></tr>
<tr><td>• Land</td><td>3.5 sq mi (9.1 km^2)</td></tr>
<tr><td>• Water</td><td>0.1 sq mi (0.2 km^2)</td></tr>
<tr><td>Elevation</td><td>902 ft (275 m)</td></tr>
</table>

Population (2010)	
• Total	9405
• Density	1357.0/sq mi (523.9/km^2)
Time zone	Eastern (EST) (UTC-5)
• Summer (DST)	EDT (UTC-4)
ZIP code	47396
Area code(s)	765
FIPS code	18-86084[1]
GNIS feature ID	0446404[2]
Website	[3]

Yorktown is a town in Mount Pleasant Township, Delaware County, Indiana, United States. The population was 9,405 at the 2010 census. It is part of the Muncie, IN Metropolitan Statistical Area.

Geography

According to the United States Census Bureau, the town has a total area of 3.6 square miles (9.3 km^2), of which, 3.5 square miles (9.1 km^2) of it is land and 0.1 square miles (0.26 km^2) of it (1.67%) is water.

Demographics

As of the census[1] of 2000, there were 4,785 people, 1,842 households, and 1,368 families residing in the town. The population density was 1,357.0 people per square mile (523.4/km²). There were 1,940 housing units at an average density of 550.2 per square mile (212.2/km²). The racial makeup of the town was 97.64% White, 0.96% African American, 0.15% Native American, 0.36% Asian, 0.23% from other races, and 0.67% from two or more races. Hispanic or Latino of any race were 0.73% of the population.

There were 1,842 households out of which 37.5% had children under the age of 18 living with them, 61.5% were married couples living together, 10.2% had a female householder with no husband present, and 25.7% were non-families. 21.8% of all households were made up of individuals and 8.6% had someone living alone who was 65 years of age or older. The average household size was 2.53 and the average family size was 2.96.

In the town the population was spread out with 26.7% under the age of 18, 7.2% from 18 to 24, 29.0% from 25 to 44, 23.6% from 45 to 64, and 13.5% who were 65 years of age or older. The median age was 37 years. For every 100 females there were 91.4 males. For every 100 females age 18 and over, there were 87.4 males.

The median income for a household in the town was $50,974, and the median income for a family was $58,784. Males had a median income of $41,346 versus $26,611 for females. The per capita income for the town was $26,065. About 3.9% of families and 4.1% of the population were below the poverty line, including 6.0% of those under age 18 and 4.3% of those age 65 or over.

History

During the Woodland period Native Americans built an earthen enclosure just to the east of Yorktown, still visible on Google Earth at 40°10′50″N 85°28′10″W. Yorktown lies at the junction of the White River and Buck Creek. According to local legend, the Miami Indians believed that the peculiar configuration of the junction made Yorktown immune from tornadoes.

Yorktown was platted in 1837 by Oliver H. Smith, who represented Indiana in the U.S. Senate from 1837 to 1843 and was a member of the Committee on Public Lands. Smith eventually became involved in the railroad business, and Yorktown was joined to Indianapolis by railroad in the early 1850s. Yorktown's main street bears Smith's name.[4]

Yorktown benefited from the 1880s natural gas boom in the area, and was the site of several glass factories in the late nineteenth and early twentieth centuries. (The gas gave names to nearby the towns of Gaston and Gas City and drew the Ball Brothers to Muncie.)

In 1892, a developer platted "West Muncie" on land immediately adjacent to Yorktown (but roughly ten miles from Muncie.) Buck Creek was dammed to form "Lake Delaware," which became the focus of a 73-room resort hotel opened in 1893. However, the dam broke within a few years and the entire West Muncie project was abandoned.[5] [6] Its most enduring legacy was perhaps that Yorktown was erroneously labeled "West Muncie" on some road maps into the 1960s and perhaps later, puzzling most local residents, who had neither seen nor heard the name in any other context.

The town has been served by the Big Four Railroad and its successors: the New York Central, Penn Central, Conrail, and CSX. The town was also served by an electric interurban line, the Union Traction Company of Indiana and its successor Indiana Railroad, in the early twentieth century.

In the mid twentieth century, many residents found employment in automotive plants in nearby Muncie and Anderson, most associated with General Motors. General Motors in Yorktown closed down in 2003. Borg Warner in Yorktown closed in 2009. Yorktown also served as corporate headquarters of the Marsh Supermarkets chain from 1952 until 1991, a fact reflected in the chain's "Yorktown" store brand. Marsh Supermarkets and Village Pantry headquarters are now located in Indianapolis, Indiana.[7]

Notes

[1] "American FactFinder" (http://factfinder.census.gov). United States Census Bureau. . Retrieved 2008-01-31.

[2] "US Board on Geographic Names" (http://geonames.usgs.gov). United States Geological Survey. 2007-10-25. . Retrieved 2008-01-31.

[3] http://www.yorktownindiana.org/

[4] Lasley, Norma of Delaware County Historical Society (2007-03-31). "Correspondence".

[5] unknown (1985-01-27). "Journal Album". Muncie Star.

[6] unknown (1898-03-24). "unknown". Muncie Morning News.

[7] "Marsh corporate website" (http://web.archive.org/web/20070207030909/http://www.marsh.net/mtc_chis.html). Archived from the original (http://www.marsh.net/mtc_chis.html) on 2007-02-07. . Retrieved 2007-02-17.

Notable Yorktownians

- Ted Haggard, prominent Evangelical Christian, grew up in Yorktown.
- James Armstrong, former President of the National Council of Churches, spent his infancy in Yorktown.
- Jacob Bartlett, writer/artist on the webcomic Terminal Crossing, grew up in Yorktown.
- Jonathon Newby, founding member of music group Brazil grew up in Yorktown.
- Carl Storie, Lead singer of the band Faith Band

Education

Elementary schools

- Pleasant View Elementary (Grade K-2)
- Yorktown Elementary School (Grades 3-5)

Middle schools

- Yorktown Middle School (Grades 6-8)

High schools

- Yorktown High School

Alternative Schools

- Youth Opportunity Center

Pro sports

- Revolution Football (http://www.revolutionfootball.us)

External links

- Town of Yorktown, Indiana website (http://www.yorktownindiana.org/)
- Revolution Football Semi-Pro Team of Yorktown, Indiana website (http://www.revolutionfootball.us/)
- Chamber of Commerce (http://www.yorktowninchamber.org/)
- Yorktown News at Muncie Free Press (http://www.munciefreepress.com/news/tag/Yorktown)

Delaware Township, Delaware County, Indiana

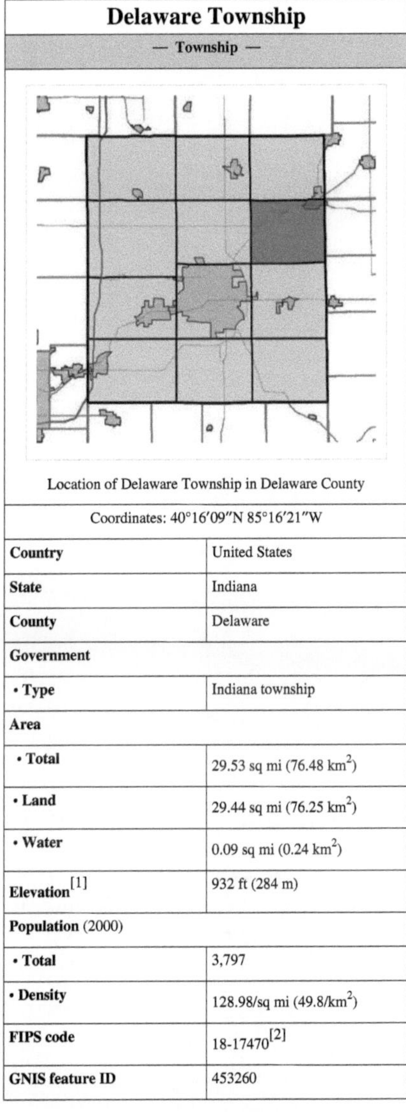

	Delaware Township
	— Township —

Location of Delaware Township in Delaware County

Coordinates: 40°16′09″N 85°16′21″W

Country	United States
State	Indiana
County	Delaware
Government	
• Type	Indiana township
Area	
• Total	29.53 sq mi (76.48 km^2)
• Land	29.44 sq mi (76.25 km^2)
• Water	0.09 sq mi (0.24 km^2)
Elevation[1]	932 ft (284 m)
Population (2000)	
• Total	3,797
• Density	128.98/sq mi (49.8/km^2)
FIPS code	18-17470[2]
GNIS feature ID	453260

Delaware Township is one of twelve townships in Delaware County, Indiana. As of the 2000 census, its population was 3,797.

Geography

Delaware Township covers an area of 29.53 square miles (76.5 km^2); 0.09 square miles (0.23 km^2) (0.3 percent) of this is water.

Cities and towns

* Albany (south half)

Unincorporated towns

* DeSoto

Adjacent townships

* Niles Township (north)
* Richland Township, Jay County (northeast)
* Green Township, Randolph County (east)
* Monroe Township, Randolph County (southeast)
* Liberty Township (south)
* Center Township (southwest)
* Hamilton Township (west)
* Union Township (northwest)

Major highways

* **28** Indiana State Road 28
* **67** Indiana State Road 67
* **167** Indiana State Road 167

Cemeteries

The township contains four cemeteries: Black, Godlove, Strong and Union.

References

* "Delaware Township, Delaware County, Indiana" [3]. Geographic Names Information System, U.S. Geological Survey. Retrieved 2009-09-24.
* United States Census Bureau cartographic boundary files [4]

[1] "US Board on Geographic Names" (http://geonames.usgs.gov). United States Geological Survey. 2007-10-25. . Retrieved 2008-01-31.

[2] "American FactFinder" (http://factfinder.census.gov). United States Census Bureau. . Retrieved 2008-01-31.

[3] http://geonames.usgs.gov/pls/gnispublic/f?p=gnispq:3:::NO::P3_FID:453260

External links

* Indiana Township Association (http://www.indianatownshipassoc.org/)
* United Township Association of Indiana (http://unitedtownships.org/)

Hamilton Township, Delaware County, Indiana

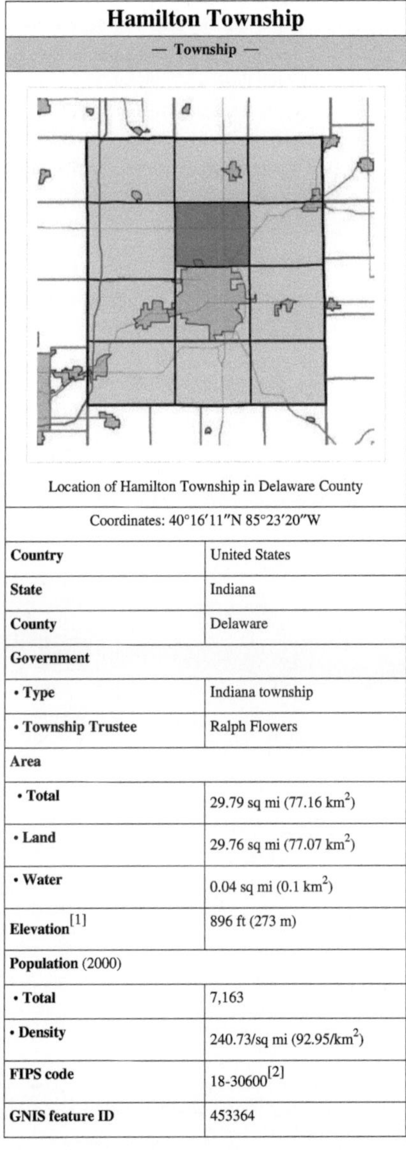

Hamilton Township	
— Township —	

Location of Hamilton Township in Delaware County

Coordinates: 40°16′11″N 85°23′20″W

Country	United States
State	Indiana
County	Delaware
Government	
• Type	Indiana township
• Township Trustee	Ralph Flowers
Area	
• Total	29.79 sq mi (77.16 km^2)
• Land	29.76 sq mi (77.07 km^2)
• Water	0.04 sq mi (0.1 km^2)
Elevation[1]	896 ft (273 m)
Population (2000)	
• Total	7,163
• Density	240.73/sq mi (92.95/km^2)
FIPS code	18-30600[2]
GNIS feature ID	453364

Hamilton Township is one of twelve townships in Delaware County, Indiana. As of the 2000 census, its population was 7,163.

Fire Department

Hamilton Township Vol. Fire Department is currently the only working fire department in the township. There are 30+ firefighters that work all for the same cause that is to preserve life and property in the township. The department currently operates 9 trucks ranging from medical and utility trucks all the way to specialised pumpers and tankers. On Sept. 12th 2009 the department hosted its yearly hog roast fund raiser for the community to raise funds that help operate the fire department all year long. This year there was 32 whole hogs cooked and over 3,000 people served. You can go to http://www.htvfcin.org to visit the Hamilton Township Vol. Fire Department's web site or call (765)282-4041 for any information or needs that the fire department can help with. Also in the spring time every year the department fills pools and hot tubs in the township. Prices on pool fills vary by size so please call for pricing and to schedule an appointment to have your pool filled.

Geography

Hamilton Township covers an area of 29.79 square miles (77.2 km^2); 0.04 square miles (0.10 km^2) (0.13 percent) of this is water.

Cities and towns

- Muncie (north edge)

Unincorporated towns

- Anthony
- Royerton

Adjacent townships

- Union Township (north)
- Niles Township (northeast)
- Delaware Township (east)
- Liberty Township (southeast)
- Center Township (south)
- Harrison Township (west)
- Washington Township (northwest)

Major highways

- 35 U.S. Route 35
- 3 Indiana State Road 3
- 28 Indiana State Road 28
- 67 Indiana State Road 67

Cemeteries

The township contains two cemeteries: Cullen and Gardens of Memory.

References

- "Hamilton Township, Delaware County, Indiana" [3]. Geographic Names Information System, U.S. Geological Survey. Retrieved 2009-09-24.
- United States Census Bureau cartographic boundary files [4]

[1] "US Board on Geographic Names" (http://geonames.usgs.gov). United States Geological Survey. 2007-10-25. . Retrieved 2008-01-31.

[2] "American FactFinder" (http://factfinder.census.gov). United States Census Bureau. . Retrieved 2008-01-31.

[3] http://geonames.usgs.gov/pls/gnispublic/f?p=gnispq:3:::NO::P3_FID:453364

External links

- Indiana Township Association (http://www.indianatownshipassoc.org/)
- United Township Association of Indiana (http://unitedtownships.org/)

Harrison Township, Delaware County, Indiana

<table>
<tr><td colspan="2" align="center">Harrison Township
— Township —</td></tr>
<tr><td colspan="2">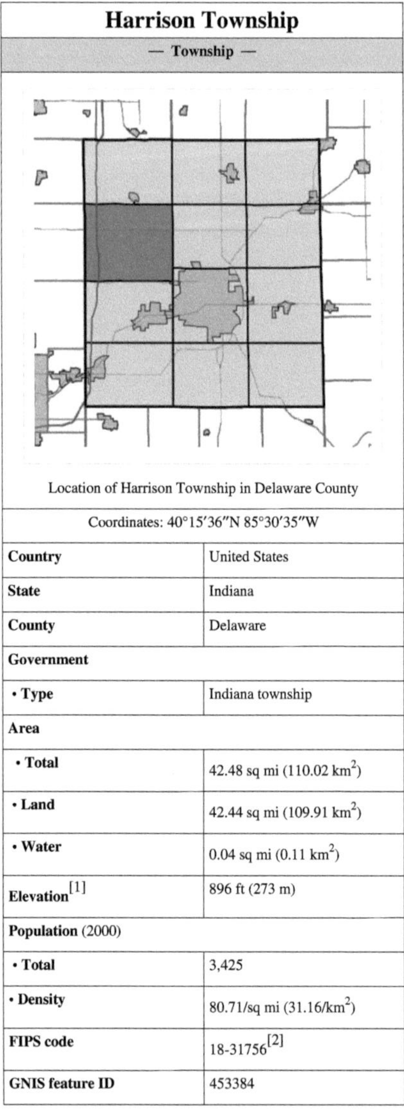
Location of Harrison Township in Delaware County</td></tr>
<tr><td colspan="2" align="center">Coordinates: 40°15′36″N 85°30′35″W</td></tr>
<tr><td>Country</td><td>United States</td></tr>
<tr><td>State</td><td>Indiana</td></tr>
<tr><td>County</td><td>Delaware</td></tr>
<tr><td>Government</td><td></td></tr>
<tr><td>• Type</td><td>Indiana township</td></tr>
<tr><td>Area</td><td></td></tr>
<tr><td>• Total</td><td>42.48 sq mi (110.02 km^2)</td></tr>
<tr><td>• Land</td><td>42.44 sq mi (109.91 km^2)</td></tr>
<tr><td>• Water</td><td>0.04 sq mi (0.11 km^2)</td></tr>
<tr><td>Elevation[1]</td><td>896 ft (273 m)</td></tr>
<tr><td>Population (2000)</td><td></td></tr>
<tr><td>• Total</td><td>3,425</td></tr>
<tr><td>• Density</td><td>80.71/sq mi (31.16/km^2)</td></tr>
<tr><td>FIPS code</td><td>18-31756[2]</td></tr>
<tr><td>GNIS feature ID</td><td>453384</td></tr>
</table>

Harrison Township is one of twelve townships in Delaware County, Indiana. As of the 2000 census, its population was 3,425.

Geography

Harrison Township covers an area of 42.48 square miles (110.0 km^2); 0.04 square miles (0.10 km^2) (0.09 percent) of this is water. Emerald Lake is in this township.

Unincorporated towns

- Bethel

Adjacent townships

- Washington Township (north)
- Union Township (northeast)
- Hamilton Township (east)
- Center Township (southeast)
- Mount Pleasant Township (south)
- Richland Township, Madison County (southwest)
- Monroe Township, Madison County (west)
- Van Buren Township, Madison County (northwest)

Major highways

- Interstate 69
- U.S. Route 35
- Indiana State Road 28
- Indiana State Road 332

Cemeteries

The township contains one cemetery, Nottingham.

References

- "Harrison Township, Delaware County, Indiana" [3]. Geographic Names Information System, U.S. Geological Survey. Retrieved 2009-09-24.
- United States Census Bureau cartographic boundary files [4]

[1] "US Board on Geographic Names" (http://geonames.usgs.gov). United States Geological Survey. 2007-10-25. . Retrieved 2008-01-31.

[2] "American FactFinder" (http://factfinder.census.gov). United States Census Bureau. . Retrieved 2008-01-31.

[3] http://geonames.usgs.gov/pls/gnispublic/f?p=gnispq:3:::NO::P3_FID:453384

External links

- Indiana Township Association (http://www.indianatownshipassoc.org/)
- United Township Association of Indiana (http://unitedtownships.org/)

Article Sources and Contributors

Selma, Indiana *Source*: http://en.wikipedia.org/w/index.php?title=Selma%2C_Indiana *Contributors*: Danorton, Grmmamma, Huwmanbeing, Ixnayonthetimmay, Jllm06, John Cardinal, Marvin01, Omnedon, Ram-Man, Robertjohnsonrj, Tr1290, Tysto, 18 anonymous edits

Liberty Township, Delaware County, Indiana *Source*: http://en.wikipedia.org/w/index.php?title=Liberty_Township%2C_Delaware_County%2C_Indiana *Contributors*: Hmains, Huwmanbeing, Jllm06, Omnedon

Delaware County, Indiana *Source*: http://en.wikipedia.org/w/index.php?title=Delaware_County%2C_Indiana *Contributors*: AshyLarry, Barneca, Brentoli, CanisRufus, Charles Edward, Chris24, D6, Drake&Josh1, George Burgess, Gman2337, GrahamHardy, Harrahs1, IsWayneBradygonnahavetosmackabitch103, Johnpacklambert, Jonathan.s.kt, Lano'reillygid, Leslie Mateus, Lord Pistachio, M dorothy, Manishv, Mlread, Monegasque, Munciedowntown, Numbo3, Nyttend, Omnedon, Owen, Owlho0t3, Pearle, Ram-Man, Rhatsa26X, Robertjohnsonrj, Ryan524, Smallbones, Sortior, Template namespace initialisation script, Tim!, TravisTX, Tysto, Utopies, Vrysxy, WhisperToMe, WillC, Woohookitty, Worldenc, 17 anonymous edits

Muncie, Indiana *Source*: http://en.wikipedia.org/w/index.php?title=Muncie%2C_Indiana *Contributors*: (jarbarf), A930913, Abeunderhill, Acntx, Ahoerstemeier, Alexf, Andrew42280, Arda Xi, Atty18, BUnderwood45, Bdean42, BernardBailyn, Bill "Hemingray" Meier, Billy Hathorn, Bluerasberry, Bob Burkhardt, Butwhatdoiknow, C.lettingaAV, CJLippert, Caltas, Can't sleep, clown will eat me, CaribDigita, Chairman Meow, Charles Edward, Chris4682, Cmr08, Ctabari11, DIDouglass, DanTD, Davodd, Deflective, Delirium, Dellkyel, DennyColt, Discospinster, Doc Strange, Dorftrottel, Drmies, Dsljrg98, Dubchubbub, Dudeman5685, ERcheck, El T, EncMstr, Erich Schneider, Ermanon, Erpbridge, Everyking, Fasach Nua, Felipito1818, Fj280, Flamerod watersoftener, Franklinandbash132, Fratrep, Frazzydee, Fryed-peach, Funky Monkey, GUllman, Gaycowboy12, Gene93k, GoCuse44, Gwernol, Harpchad, Hephaestos, Heyjude1971, Hi540, Histrion, Hit bull, win steak, HowardBerry, Hraefen, Huwmanbeing, IKato, Ichabod, Igordebraga, Iohannes Animosus, Ixnayonthetimmay, JDNeckers, JHawk88, Jackola, Jaosmith, Jasonbuie, Jcurtis, Jdbrendel, Jeepday, Jeremiestrother, Jllm06, Jnelson09, Joelvanatta, John Cardinal, Jonathan Versen, Joshualross, JustAGal, Justin W Smith, Katherine, Kbdank71, Kbh3rd, Kilonum, Kittybrewster, Klausness, Korg, Kpaul.mallasch, KryptoCleric, Kuru, Kwamikagami, L Kensington, L., Letitgotopperthekingisgone, Lincolnite, Lord Pistachio, Lurleen, MJCdetroit, Madmanblake, Malik Shabazz, MarB4, Martin Osterman, Marvin01, Maxbrando2000, Mellery, Mervyn, Mike Rosoft, Mincebert, MisfitToys, Mithridates, Mlaffs, Mlread, Moeron, Moorematthews, MrChupon, Mrschimpf, Msdavey, Mszajewski, MuncieIndiana, Muncieauthority, Munciedowntown, Munciegadgetguy, Nakon, Namikiw, Natalie Erin, NawlinWiki, Nolesrock, OckRaz, Ohconfucius, Olovni, Omnedon, Otolemur crassicaudatus, PBS-AWB, PIrish, Packerfansam, Pacobob, PartTimePvnk, Patleahy, Paul Soth, Phaedriel, PhilHibbs, Prairiefarmer, Pwkaskie, R'n'B, RainbowOfLight, Ram-Man, RayvnEQ, Readynoise07, Redsox00002, RekishiEJ, Rhatsa26X, Rich Farmbrough, RichardF, Rjensen, Robbiediva, Robertjohnsonrj, SarahStierch, SarekOfVulcan, Searchme, Sedna10387, Seth Ilys, Shagreg, Sillybillypiggy, Slambo, SoCalSuperEagle, Sophie, Squirrley, Steel, Stickee, Strongfortress, Stupid104, TJRC, TRBP, Tassedethe, Template namespace initialisation script, TenPoundHammer, Tesi1700, The Rambling Man, Themrwilk, Togieharold, Tortillovsky, Tysto, Ulric1313, Uris, Vdoubleuman, Vladimir Koloff, WODUP, Whosyourjudas, Wikiwatch, William Graham, WilliamThweatt, Xiahou, Yuckfoo, 551 anonymous edits

Albany, Indiana *Source*: http://en.wikipedia.org/w/index.php?title=Albany%2C_Indiana *Contributors*: Acntx, AlanaKylie, Amasters321, Bazonka, Ben Ben, Harpchad, Ixnayonthetimmay, Jimithing55, Jllm06, John Cardinal, MJCdetroit, Malepheasant, Marvin01, Mellery, Nem80, Omnedon, Pearle, Ram-Man, Rettetast, Robertjohnsonrj, SUL, Tremont798, Tysto, Wp bh726, 11 anonymous edits

Chesterfield, Indiana *Source*: http://en.wikipedia.org/w/index.php?title=Chesterfield%2C_Indiana *Contributors*: Geoking66, Harpchad, Jeremiestrother, Jllm06, John Cardinal, MJCdetroit, Marvin01, Mellery, Nprater, Nyttend, Omnedon, Pearle, Ram-Man, Robertjohnsonrj, Sedna10387, TheHoosierState89, Tysto, 6 anonymous edits

Daleville, Indiana *Source*: http://en.wikipedia.org/w/index.php?title=Daleville%2C_Indiana *Contributors*: CommonsDelinker, Droll, Harpchad, Huwmanbeing, Ixnayonthetimmay, Jllm06, Jmiller646, MJCdetroit, Marvin01, Mellery, Omnedon, Pearle, Philip Trueman, Ram-Man, Robertjohnsonrj, Thesoundguy, Tysto, 12 anonymous edits

Eaton, Indiana *Source*: http://en.wikipedia.org/w/index.php?title=Eaton%2C_Indiana *Contributors*: Charles Edward, Chris the speller, CommonsDelinker, Deltarocks, Harpchad, Ixnayonthetimmay, Jllm06, John Cardinal, LilHelpa, MJCdetroit, Malepheasant, Marvin01, Mellery, Nyttend, Omnedon, Pearle, Ram-Man, Robertjohnsonrj, Tysto, 5 anonymous edits

Gaston, Indiana *Source*: http://en.wikipedia.org/w/index.php?title=Gaston%2C_Indiana *Contributors*: CommonsDelinker, Harpchad, Ixnayonthetimmay, Jllm06, John Cardinal, MJCdetroit, Marvin01, Mellery, Omnedon, Pearle, Ram-Man, Robertjohnsonrj, Tysto, 3 anonymous edits

Yorktown, Indiana *Source*: http://en.wikipedia.org/w/index.php?title=Yorktown%2C_Indiana *Contributors*: AndrewWeaver2, Can't sleep, clown will eat me, Chris the speller, Funwithbig, FusionNow, Huwmanbeing, Ixnayonthetimmay, Jack22782, JamesAM, Jllm06, Joelvanatta, Kvdveer, Marvin01, Moe Epsilon, Omnedon, Pigsonthewing, Ram-Man, Rexdwyer, Robertjohnsonrj, Sedna10387, Tysto, Ulric1313, WilliamThweatt, Xcentaur, 38 anonymous edits

Delaware Township, Delaware County, Indiana *Source*: http://en.wikipedia.org/w/index.php?title=Delaware_Township%2C_Delaware_County%2C_Indiana *Contributors*: Hmains, Huwmanbeing, Jllm06, Omnedon

Hamilton Township, Delaware County, Indiana *Source*: http://en.wikipedia.org/w/index.php?title=Hamilton_Township%2C_Delaware_County%2C_Indiana *Contributors*: EoGuy, Hmains, Huwmanbeing, Jllm06, Omnedon, 5 anonymous edits

Harrison Township, Delaware County, Indiana *Source*: http://en.wikipedia.org/w/index.php?title=Harrison_Township%2C_Delaware_County%2C_Indiana *Contributors*: Hmains, Huwmanbeing, Jllm06, Omnedon, Ulric1313

Image Sources, Licenses and Contributors

Printed by Books on Demand GmbH, Norderstedt / Germany